Mouse-eared bat

Virginia opossum

Pygmy marmoset

Snowshoe hare

Spotted skunk

Golden mouse

Fairy armadillo

Eastern chipmunk

Cottontail rabbit

Common dormouse

Creatures
Small and Furry

by Donald J. Crump

A chipmunk with its cheeks full of seeds peeks out of a hole.

☐ BOOKS FOR YOUNG EXPLORERS
NATIONAL GEOGRAPHIC SOCIETY

Small furry creatures live all
around us, near our homes
and in faraway places.
They are often hard to see.
Some hide under the ground.
Others climb high in trees.
Many come out only at night.

As darkness falls, three little
dormice creep out of their nest.
A dormouse is so small that
it would easily fit
in your hand.

A mother dormouse curls up with her babies in their grassy nest. She feeds her young with her milk. Then she licks them clean with her pink tongue. Soon they will be old enough to go out on their own.

Four young opossums hold on
to their mother's thick fur.
She carries her young with her
as she goes out looking for food.

A mother ground squirrel nurses
her baby with milk from her body.
Animals that have fur or hair and
nurse their young are mammals.

As snow falls, a pine marten waits and watches on an icy log. A thick winter coat helps keep the marten warm and dry.

Fur can also protect a mammal from enemies. The fur of the snowshoe hare turns white in winter. The hare can hide in the snow. By summer, its fur has become brown like the ground. This hare's ears, nose, and feet will soon be brown, too.

How do you think this cotton-top tamarin got its name? On its head, it has a tuft of long white hairs. Its face has a rough beard.

All mammals have hair, but they do not all look alike.
This fat little dormouse has long outer hairs and short thick
fur underneath. Do you see its long whiskers? The fairy
armadillo digging in the dirt has almost no hair at all.
It has a few hairs around its hard shell and on its belly.

These mammals are grooming themselves. This helps keep their hair and skin clean.

A cottontail rabbit licks a paw. It cleans its fluffy fur with the wet paw. This little mouse begins a bath by combing its whiskers with two tiny paws.

Two striped chipmunks sip water from a forest pool. You can find chipmunks in woods, in parks, and maybe even in your backyard.

The pygmy marmoset is the
smallest monkey. This marmoset
on a jungle vine is eating
a grasshopper. As it leaps
among the trees, the marmoset
chirps, "Gee, gee, gee."

A vole nibbles on a blade of grass.
This cousin of the mouse eats
plants in fields and gardens.

Water shrews live near streams and lakes. They spend most of their time hunting for food and eating.

These animals have hairy feet that help them swim. One shrew dives under the water and pokes its nose among the rocks. Soon the shrew will rise to the surface of the water to breathe. Another shrew swims along the surface, then climbs onto a rock.

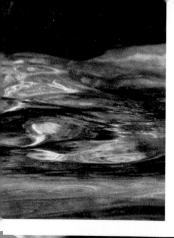

At dusk, wide-eyed flying squirrels come out and look for food. They glide from tree to tree. A flying squirrel does not have wings. The skin between its arms and legs acts like a parachute. The arms and tail help the squirrel steer.

Furry creatures have their own ways of talking to each other.
A playful young marmot nuzzles its mother. Marmots are
the largest of the squirrels. They live in family groups.

EEK! A short-tailed shrew opens its mouth wide and screeches.
It is warning another shrew to stay away from its nest.
These little animals are fighters and usually live alone.

With a bobcat close behind, a mouse dashes across the snow.
How do you think small creatures defend themselves?

A skunk stands on its front paws and aims a stinky spray at an
enemy. A silky anteater holds its sharp claws in front of its face.

Bats are the only mammals that
really fly. With wings stretched
wide, a bat flits about in the dark.
This bat catches insects to eat.

Kangaroo rats hop about and fight each other with strong hind legs. Their long tails help them keep their balance. Hippety-hop! With long leaps, a cottontail rabbit races away. It runs and dodges when other animals chase it.

Under the cornstalks, a
cottontail rabbit is hiding.
It sits very still in the snow.
Enemies will have a
hard time seeing it.

Another kind of rabbit digs tunnels for secret hiding places.
Many rabbits share this underground home. How many do
you see? One rabbit pops out for a look around the meadow.
In a tunnel, a mother rabbit is caring for her young. She built
a nest for them with dry plants and her own soft fur.

All around you, little furry creatures run and hide. Deep in a wheat field, two harvest mice meet whisker to whisker. What do you think they would say to each other?

Published by
The National Geographic Society
Gilbert M. Grosvenor, *President*
Melvin M. Payne, *Chairman of the Board*
Owen R. Anderson, *Executive Vice President*
Robert L. Breeden, *Vice President,*
    Publications and Educational Media

Prepared by
The Special Publications Division
Donald J. Crump, *Director*
Philip B. Silcott, *Associate Director*
William L. Allen, William R. Gray,
    Assistant Directors

Staff for this book
Jane H. Buxton, *Managing Editor*
Alison Wilbur Eskildsen, *Picture Editor*
Cinda Rose, *Art Director*
Peggy D. Winston, *Researcher*
Patricia Larkin, *Assistant Researcher*
Rebecca Bittle Johns, Carol A. Rocheleau,
    Illustrations Assistants

Nancy F. Berry, Pamela A. Black, Mary Frances Brennan,
    Mary Elizabeth Davis, Rosamund Garner,
    Victoria D. Garrett, Virginia A. McCoy, Cleo E. Petroff,
    Tammy Presley, Sheryl A. Prohovich, Kathleen T. Shea,
    Staff Assistants

Engraving, Printing, and Product Manufacture
Robert W. Messer, *Manager*
George V. White, *Production Manager*
Mary A. Bennett, *Production Project Manager*
Mark R. Dunlevy, Richard A. McClure,
    Gregory Storer, *Assistant Production Managers*
Katherine H. Donohue, *Senior Production Assistant*
Julia F. Warner, *Production Staff Assistant*

Consultants
Dr. Glenn O. Blough, *Educational Consultant*
Lynda Ehrlich, *Reading Consultant*
Dr. Margaret A. O'Connell, National Zoological Park,
    Scientific Consultant

Illustrations Credits
Owen Newman/NATURE PHOTOGRAPHERS LTD. (cover, 2-3, 4-5, 10-11); ANIMALS ANIMALS/John Gerlach (1); Rick McIntyre (6); Richard S. Diego (6-7); Harry Engels (8, 12-13, 22-23); Greg Beaumont (9); Rod Williams/BRUCE COLEMAN LTD. (10 upper); Tony Morrison (10 lower, 24 right); William J. Weber (13); Joseph Van Wormer/BRUCE COLEMAN LTD. (14-15); Wolfgang Bayer (16); G. I. Bernard/OXFORD SCIENTIFIC FILMS (16-17); Dwight R. Kuhn (18-19 upper, 23 lower); ANIMALS ANIMALS/Stouffer Enterprises Inc. (18, 19, 24-25); Jane Burton/BRUCE COLEMAN LTD. (18-19 lower); Hans Reinhard/BRUCE COLEMAN LTD. (26-27 upper); Willis Peterson (27); Tom Brakefield (28); ANIMALS ANIMALS/Oxford Scientific Films (30-31); Wayne Lankinen/BRUCE COLEMAN INC. (32).

All artwork by Robert Hynes. Painting on page 20-21 based on photographs by J. Sherwood Chalmers/Greg Dale (flight) and William J. Weber (feeding).

Library of Congress CIP Data

Crump, Donald J., 1929-
    Creatures small and furry.

    (Books for young explorers)
    Summary: Text and photos present characteristics, habits, and habitats of small and furry mammals, including the dormouse, pine marten, cotton-top tamarin, pygmy marmoset, and water shrew.
    1. Mammals—Juvenile literature. [1. Mammals] I. Title. II. Series.
QL706.2.C78 1983   599   83-13456
ISBN 0-87044-486-7 (regular edition)
ISBN 0-87044-491-3 (library edition)

Tail in the air and feet tucked under,
a chipmunk heads out of sight.

Cover: A harvest mouse weighs so little
that it hardly bends a stalk of wheat.

Water shrew

Cotton-top tamarin

Bannertail kangaroo rat

European rabbit

Arctic ground squirrel

Harvest mouse

Deer mouse

Silky anteater

Hoary marmot

American marten

Short–tailed vole

Flying squirrel

The smallest of these animals would fit in your hand. The biggest is about the size of a house cat.